the guide to owning
Koi

MW00522629

David E. Boruchowitz

T.F.H. Publications, Inc.
One TFH Plaza
Third and Union Avenues
Neptune City, NJ 07753

Copyright © 2002 by T.F.H. Publications, Inc.

All rights reserved. No part of this publication may be reproduced, stored, or transmitted in any form, or by any means electronic, mechanical or otherwise, without written permission from T.F.H. Publications, except where permitted by law. Requests for permission or further information should be directed to the above address.

This book has been published with the intent to provide accurate and authoritative information in regard to the subject matter within. While every precaution has been taken in preparation of this book, the publisher and author assume no responsibility for errors or omissions. Neither is any liability assumed for damages resulting from the use of the information herein.

ISBN 0-7938-0373-X

If you purchased this book without a cover you should be aware that this book is stolen. It was reported as unsold and destroyed to the publisher and neither the author nor the publisher has received any payment for this "stripped book."

Printed and bound in the United States of America

Printed and Distributed by T.F.H. Publications, Inc.
Neptune City, NJ

Contents

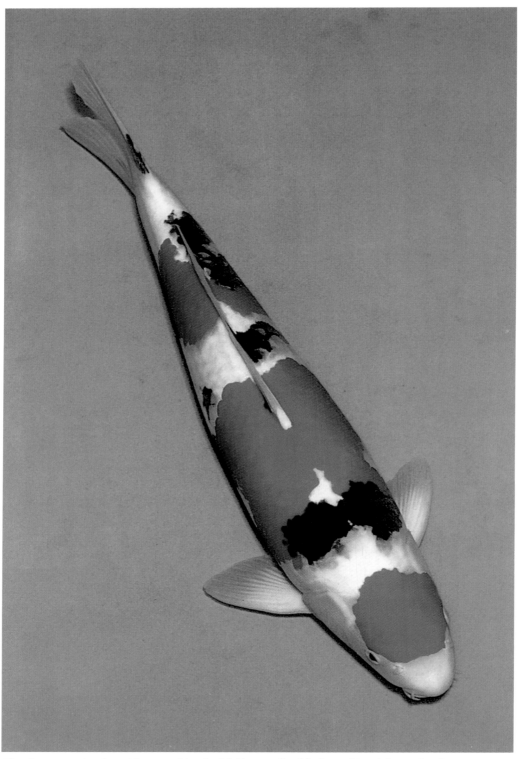

The three round red markings combined with fine-quality black on this taisho sanke champion are a koi enthusiast's definition of perfect.

Koi and Koi Ponds

There is probably no other fish more closely associated with garden ponds than the magnificent colored carp the Japanese call *nishikigoi*. Although their shape and habits still identify them as carp (*Cyprinus carpio*), it is hard to believe that these brilliantly colored fish are even related to the drab, dull green-brown ancestral fish. Anyone used to the "trash" nature of common carp would be flabbergasted at the five- and even six-figure sums of money a single prize-winning show koi can command! I just took a quick look at several Internet sites advertising young breeding stock for sale, and almost all were at least $1000 for a single adult koi.

Fortunately for the average pond-keeper, the culls from prize-winning stock are still very beautiful fish, and they can be purchased much more reasonably. Keeping and raising these gorgeous fish is not a difficult task, and a koi pond can be an exquisite addition to your garden.

There are certain minimum requirements for keeping koi, however, so you should familiarize yourself with the proper care of koi before deciding if they are for you. In this book we will look at the koi themselves and at their proper care. You will read about how to house, feed, and care for these wonderful animals.

THE POND

Can koi be kept in aquaria? Not really. Oh, a single koi could live in a 100-gallon tank for a while—their full adult size is about 3 feet, with the occasional 4-foot individual. But koi are best kept in schools, they have been bred for centuries to be viewed from above, and they are large, hardy, active fish. It is therefore in ponds that they are best kept and best viewed. This does not mean that koi must be kept outside; in fact, in Japan indoor koi pools are common. Some

This lovely fish is called an asagi. The scale arrangement is exquisite and the little touches of red are like punctuation marks.

are even indoor-outdoor ponds, so that the koi can swim from garden to living room, and their owners can appreciate them inside and out. It is, however, in large outdoor garden ponds that koi are most easily appreciated and most often found.

FILTRATION

The first consideration for a koi pond is filtration. Why? Well, koi are phenomenal waste generators. Heavy-bodied and heavy-feeding, koi produce prodigious amounts of ammonia and other wastes. This indicates a need for a very efficient biofilter. In addition, koi produce a lot of solid waste, and, being bottom-sifting scavengers, they also stir up a lot of sediment—hence the need for heavy-duty mechanical filtration. The facts that koi look good in the sun, plus the amount of nutrients they produce, plus the fact that they can be hard on nutrient-absorbing plants, all mean that pea soup algae conditions are quite likely in the koi pond—hence the widespread use of UV sterilizers.

"The bigger, the better" is the best answer to the question of how large a filter you will need. Most commercial garden pond filters are woefully inadequate except as prefilters to your main unit. The ideal koi pond filter is built integrally with the pond. It costs very little extra to pour a little more concrete or place a few more square yards of pond liner at the time of the original installation, and at this point the plumbing is a breeze. A series of chambers is a common design, with various filter materials being used.

Water Scrubbing

The water should first pass through mechanical filtration media designed to trap and hold suspended wastes and dirt. Filter brushes, pads, or fibers can be used, alone or in combination. A series of chambers with filter material of decreasing coarseness is very effective at trapping progressively finer suspended material.

For example, the water might be pumped from the pond into the first chamber, which is full of brushes. Plant leaves, dead insects, and large pieces of detritus will be trapped here. The next chamber could have a bed of matting or woven filter material, which would strain out the smaller bits floating around in the water. Then the water could flow into a very fine medium like filter sponge or even fine gravel, where the smallest suspended particles would be caught up, leaving only dissolved substances in the water flow. By the time you get down to dense pads or sponges, the medium is also going to be supporting a lot of biofiltration bacteria.

Biological Filtering

It is the beneficial bacteria in the biofilter that remove the toxic ammonia and other fish wastes and convert them into less harmful substances. Since the key to biofiltration is *surface area*, you need to use a medium that maximizes the area for bacterial colonization. The finest media used for mechanical filtration do supply some, but you will want at least one section of the filter dedicated just to biofiltration. Since a medium dense with microscopic pores is easily plugged, it is vital that the water is indeed scrubbed clean before entering this part of the filter. You must also clean the first sections regularly, preferably by backflushing. (A little bit of plumbing and a few valves will add little expense to the original construction of the filter but will pay off in the ease with

which backflushes can be performed.) Keeping the front compartments clean will ensure that a steady stream of clean water enters the biofiltration chamber. The biofilter medium, on the other hand, should be disturbed as little as possible, except to make sure that it is not plugged and that the water is flowing through it properly.

Various plastic, ceramic, and natural materials can be used as biofiltration

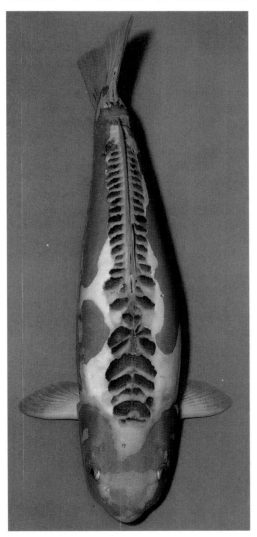

This fish is called a shusui. The dorsal scales are unique and interesting and the red coloration deepens its "personality."

media. Some people use gravel or even sand, but these media are prone to channeling—the water flow through them may not be uniform, but can follow paths of least resistance, resulting in some areas of the medium becoming anaerobic when the supply of oxygenated water is limited or even cut off. To prevent this, the medium should be gently stirred on a regular basis or a different material less prone to uneven compacting can be used. Careful design of

This is a spectacular way to return water to your pond. If water lilies are to be kept in such a pond remember they prefer more tranquil waters.

the filter baffles can also help maintain a steady flow through the biofilter.

The active bacterial colonies that strip ammonia and other dissolved wastes from the water need this unimpeded flow of oxygenated water. While fountains or cascades are a preferred method of returning water from the filter to the pond, this provides the necessary aeration after it is most needed. A very effective way of keeping the oxygen supply to the biofilter high is to locate the filter higher than the pond. The chambers of the filter are stepped, so the first section is highest, with cascading overflows between them. In such a setup the water is pumped from the pond up to the first filter chamber. It then flows by gravity through the various chambers, dropping (and aerating) from each to the next. When it exits the biofilter, it drops back into the pond.

ZAPPING ALGAE

Some green slime algae growing on rocks and on the sides of the pond does not represent a problem, and the koi will happily graze on it. Free-living unicellular algae are the cause of green water. While not unhealthy in most cases, it is unsightly, and you'll want more from your pond than just a glimpse of red every once in a while in a sea of bright green. There are three principal ways of combating unwanted algal growth.

Light

As plants, algae need light to grow, and the more light, the more algae. Of course, you want your pond plants to

UV Sterilizer Killing Algae

A UV sterilizer will kill algae and harmful pathogens.

thrive, so erecting a roof over your pond is not a good solution to the algae problem. Desirable aquatic and marginal plants need at least six hours of direct sunlight a day, but if your pond is positioned so that it is in shade or partial shade the rest of the time, they will still do well, but algal growth will not be as rampant as it would be in full sun all the time.

Nutrients

Algae are the first step in many food chains, and they are efficient users of dissolved nutrients, taking phosphates and various nitrogenous waste compounds directly from the water. Other plants, of course, also take up these nutrients. This means that one way to control algae is to get higher plants established quickly. They will use up

With the summer sun also comes "algae season." A UV sterilizer will help combat algae but so will time and a low bioload.

A heavy plant cover in the pond can help to retard the growth of algae, partly because the foliage of the plants cuts down on the sunlight available to the algae and partly because the plants compete with the algae for nutrients in the water.

nutrients as they become available, making an algal bloom unlikely.

This is easier said than done, but it is possible. It is more difficult in a koi pond simply because these fish will happily consume most plant material. It is still possible to have aquatic plants in a koi pond—you simply must take into account that some of the plants will get eaten; tougher plants like water lilies and most marginals will not be devoured by the koi.

So, how is this difficult but possible goal accomplished? We'll cover pond plants in more detail in a later chapter, but when the pond is first established, you should get a large number of quick-growing aquatic plants. These are usually called oxygenators, and while they do produce oxygen during daylight, they are even more important as nutrient sponges. They include genera such as *Elodea*, *Cabomba*,

and *Vallisneria*. Most are unrooted plants sold in bunches, with the grass-like *Vallisneria* species being the major exception. At this point you will either not have added your koi or the fish will still be small, and the plants will have a chance to grow before they are eaten.

You may be familiar with these plants as limp, 8- or 10-inch specimens in an aquarium. Well, in a pond they grow rampantly, to say the least. *Elodea* often spreads across the entire surface of the pond, blooming profusely, and pond-grown *Vallisneria* with leaves 6 feet long are common.

Meanwhile, marginal or bog plants, which grow with their roots under water and part or all of their foliage above the surface, can be established in baskets or pots along the banks of the pond. Water lilies and lotuses, planted in pots full of fertile soil, will quickly grow and provide beautiful flowers as well as nutrient uptake.

Ultraviolet Rays

Concentrated ultraviolet (UV) rays kill cells. This is why sunlight has sterilizing properties. If you supply UV radiation focused into even greater intensities and placed extremely close to microorganisms, you will have a very effective sterilization device that will kill not only algae but also protozoans, bacteria, and viruses that cause fish illnesses.

This is what UV sterilizers do. They have a bulb that gives off UV rays situated in a quartz sleeve so the water flows through in close proximity to the bulb. This will effectively kill all organisms in the water, provided three requirements are met.

1) **High Output**. The bulb must be operating at high output. When first installed, a UV bulb will operate at full intensity, but this quickly decreases, and within a matter of months the output is reduced to ineffectual levels. You do not replace these bulbs when they burn out; you must track the number of hours of operation and follow manufacturer's guidelines for replacement or even replace them sooner.

2) **Contact Time**. The contact time between the pond water and the UV must be long enough for the radiation to do its job. The water must flow through the unit in a thin stream (to maximize the intensity of the radiation) and slowly enough that it is exposed sufficiently to kill all the algae. The balance needed here is between flow rate and flow volume, or capacity.

3) **Capacity**. UV units are often rated by the number of gallons of pond water that they can handle or by the gallons per hour that can flow through. You need to maximize the amount of water that flows through the device, but you want to minimize the speed with which it does that. The volume or capacity is important because all of the water in the pond must flow through the sterilizer often enough; the low speed is necessary so that it will

Generally sold in bunches, these fast-growing aquatic plants maximize your chances of robbing the pond of algae supporting nutrients until the water lilies, lotuses, and marginals take over that function.

be exposed to the UV light long enough to be effective.

It is therefore important to use a unit of sufficient power for your setup or, even better, one rated for an even larger setup. Since koi are about the hardest fish on a filtration system, any filtration unit, including a UV sterilizer, should be rated for a pond twice the size of the one in which you plan to use it. If you have a 1500-gallon pond, get a unit stated to be sufficient for a 3000-gallon setup.

THE WATER

Fortunately, koi are quite tolerant of varying water parameters, so your water supply need not be anywhere near as complex as your filtration system. Water hardness is not crucial, as long as it is not extreme in either direction. As for pH, it is not too important in terms of the fish's health, but it does have some effect on the color of koi. The perfect realization of red coloration occurs in acidic waters, where getting good blacks is problematic. The reverse is true in alkaline waters, where black coloration is typically excellent, but it is harder to get a good red. Obviously, neutral water (pH 7) is a decent compromise.

Water Changes

A garden koi pond is not under the same pressures as an aquarium to have regular water changes, but it does need them, and its sheer size makes this necessary task more daunting. The natural environment, including beneficial sunlight and the dilution of wastes by rainfall, helps to maintain water conditions. As with any fish setup, the more water you change and the more often you change it, the happier and healthier your fish will be.

In many areas, summer means heat and not much rain. This leads to massive evaporation from the pond and a lowering of the water level. This in turn concentrates waste products in the water, since only pure water evaporates, leaving all the dissolved substances behind in a smaller volume of water. If you simply refill the pond, you miss a chance to improve water conditions. True, the wastes will be diluted by the new water, but only back to their original concentrations. If you instead drain 20 percent of the water and then refill it, you will have decreased the total amount of dissolved wastes as well as returned the water to its proper level.

Do not overestimate the amount of rain your pond receives. Many people are impressed with a furious electrical storm and downpour, but very often little water is added to a pond this way. It takes a steady, consistent rainfall to accumulate significant amounts of water. Even then, the large volume of a pond means that substantial rains can still produce very little water change.

You should use a rain gauge to determine how much water you have actually received. The "gauge" can be as simple as a pan set next to the pond. If it winds up with an inch of water in it, you can figure an inch of water fell on your pond. A quick calculation will estimate the gallons added. For example, a 10-foot by 16-foot

Big or small, formal or informal, deep or shallow, with fish or without fish, filtered or unfiltered, in temperate clime or subtropical, in the ground or above it, formed from a liner or sculpted in concrete, a garden pond is the product of its designers imagination—as is the pond's degree of simplicity or extravagance.

pond has a surface area of 160 square feet, which is 23,040 square inches. An inch of rainfall means 23,040 cubic inches of water were added to the pond, which is about 100 gallons. (One gallon is 231 cubic inches; there are 1728 cubic inches in a cubic foot.) If the pond has an average depth of 4 feet, it contains about 4800 gallons, so this represents an addition of only 2 percent new water.

This is *not* a 2 percent water change, however. If the pond overflowed, some of the water that went down the drain was the new rainwater. If the pond did not overflow, then this rainfall simply diluted the dissolved wastes with 100 gallons, but nothing was removed.

POND SIZE

Now we can consider the size of the pond itself. As with a filter, bigger is better. The various dimensions of a koi pond are all significant.

Area

There are two areas of your pond that are important—the surface area and the bottom area. In an aquarium these areas are equal, as they are in some ponds, but most ponds have sides that taper down to the bottom, meaning that the surface area is greater than the bottom area, sometimes much greater.

The surface area is important to you because that it is where you see your fish and where you grow water lilies. It is important to the koi because that is where gas exchange (aeration) takes place. The aeration of the filter and any addi-

tional devices such as fountains is extremely important, but so is the natural aeration that occurs on the water surface. Maximizing surface area maximizes the carrying capacity of your pond.

Depth

Since koi are not territorial, bottom area is not important because each fish does not stake out a decent parcel of real estate, as it is with cichlids, for example. The reason the bottom area is important to koi is related to depth.

Koi, like all carp, have a very wide range of temperature tolerance, but they have their limits. Extremely warm water will stress them if it doesn't kill them, and they cannot survive being frozen solid all winter. The depth of your pond is important to both of these extremes.

Have you ever gone swimming in a pond on a very hot summer day? Typically the first few inches at the surface are very warm water, but lower down, where your arms and legs hang, can be considerably colder. The deeper the water, the cooler the bottom will be. Your koi need to be able to escape the warm surface water and rest in cooler water at the bottom.

While what I just said is true, it is slightly misleading. As you go deeper into the water, the temperature actually becomes more stable rather than cooler. In a very deep lake, the temperature of the water at the bottom hardly varies at all day in and day out, summer and winter. In August you might enjoy swimming in the lake, and in February you might go skating on it, but at the bottom the temperature

Okay—which one of you stole my ring?

remains the same all the time. The same, of course, is true for depths that are not full of water, which is why deep caves generally stay the same temperature all the time as well.

This means that the key to your koi's survival in the winter is also the depth of your pond, since while the upper stratum freezes solid, in the deep reaches the water will remain unfrozen. The deeper your pond, the more stable the temperature will be, but since it will not be anywhere near as deep as a large lake or underground cavern, the temperature will still vary considerably, even at the bottom.

If you live in the Far North, your pond will need to be very deep to provide the koi with unfrozen water in the winter, and if you live in the Deep South, it will need to be very deep to provide them with cool water during the heat of summer. How deep?

You have to take into account the extremes of climate you normally experience, but you should also factor in for unusual weather patterns. Here in upstate New York the recommended below-frost level is 4 feet, but one winter our 6-foot-deep pond froze almost to the bottom and we had a massive koi kill. A few years later, the pond never froze solid enough to skate on! In the same way, a 4-foot depth might normally be cool enough for the koi in a Louisiana garden pond, but an unusually hot summer might turn them into bouillabaisse.

On the other hand, you're unlikely to want or to be able to provide an extremely deep excavation for your pond. One solution is to design your pond with a more reasonable overall depth but have a very deep trench at the center. In the summer your koi will head for this trench during the hottest part of the day, and in the winter they will do fine there while the rest of the pond can freeze solid. It is much easier and less expensive to build such a trench than a pond that is deep all over.

The minimum depth for a koi pond (aside from any trench) is often said to be 3 feet, but 5 feet is a better choice. Remember, once your koi are grown, they will be almost 3 feet long themselves. If your pond is 3 feet deep and you have a trench 3 feet deeper, that will give a 6-foot maximum depth, but the bulk of the pond will be a more manageable 3 feet.

Volume

Well, if you remember your math, you know that area times depth equals volume, and the volume of your pond is also important. The more water there is in your pond, the greater its capacity to hold sufficient oxygen for your fish and to absorb wastes before they reach dangerous levels. Aeration and biofiltration greatly extend that capacity, but the overall volume of the pond is significant as well. Of course, an increased pond volume is of no benefit if you couple it with an increase in the number of fish you put into it!

Stocking Rate

Koi are big fish, and they need big ponds. If you only have room for a small pond, you should stick with goldfish. You

An indoor pond is a true luxury for a koi enthusiast.

Once you're finished with the barrow, shovels, and pick-axes, the real fun of planting and stocking the pond will begin.

simply cannot crowd koi, or you will have a wealth of problems. I predict you aren't going to like my recommendation for stocking your pond, especially if you're used to the terribly inadequate aquarium cliché of an inch of fish per gallon of water. An 8-foot by 6-foot by 3-foot pond holds about 1000 gallons of water, but you cannot keep anywhere near 1000 inches of koi in it. That would be 30 full-grown koi, but that pond should ideally house only **two** large koi. That's right, figure about **25 square feet of surface** for each fish. You can stock more fish than that, but I do not recommend it.

You might argue that it takes koi a long time to grow to full size, and you don't want to look at a pond with only two or three fish in it. That is fine, as long as you are committed to thinning them out as they grow. That doesn't mean waiting until they're grown! If you start with 10-

Looking at this lovely formal pond before the filtration equipment is put into disguise gives us a good idea of the planning and level of workmanship involved. When this effort is successful the enjoyment of the result is not marred by a succession of emergencies.

inch koi, that 6 by 8 pond might handle six or eight of them comfortably. By the time they are a foot long you would definitely want no more than half a dozen. At 20 inches, you should be down to three or four fish.

There is evidence that koi that are kept in lightly stocked ponds will become sexually mature at a smaller size—as little as a foot long, while fish in more densely stocked ponds may not mature until they are about 2 feet long. Since their growth rate slows down considerably once they reach sexual maturity, the more densely stocked fish will grow faster. Of course, *over*crowded fish will grow more slowly because water quality will deteriorate, which inhibits feeding and stunts growth.

But if you avoid overcrowding, you may actually get faster growth in your koi by keeping more young fish together than will fit in the same pond as adults, as long as you plan to remove them incrementally as they get bigger to prevent overloading the system.

CONSTRUCTION

Since preformed ponds are too small for koi, most people use ponds made of concrete or with plastic or rubber liners. Natural earth ponds can be used for koi, but you will rarely see your fish. Koi cannot leave the bottom alone, and they will always be stirring it up and muddying the water.

As far as the koi are concerned, a pond liner and concrete are about equal. The flexibility, cost, and ease of construction with liners make them more popular, but concrete is often the medium of choice for ponds in more formal gardens, where perfect geometric shapes like rectangles or circles are desired. It is at the same time much stronger and less flexible. For ponds that will receive rough handling such as sharp-clawed dogs running through them, concrete is the only way to go. In areas likely to experience severe frost heaving, liners have a definite advantage.

Drains and Overflows

Installing a drain in a pond as you construct the pond is simplicity itself. Doing it as an afterthought on an established pond is a major undertaking. Only the smallest ponds, which can be easily emptied with a submersible pump, should be constructed without an adequate drain.

The drain can be as simple as a plugged opening in the bottom of the pond that you reach down to open when you wish to drain the water. You can also run a pipe from the drain with a valve to open and close the drain. In either case, the pond will need an overflow as well, so that the pond will not rise over its banks during a rainstorm. An appropriately screened pipe should be located above the normal water level but below the rim of the pond. You can also combine drain and overflow into one with a standpipe. If the drain is threaded, you can install a standpipe that terminates at the desired water level. When the water rises above this level, it will flow into the pipe and down the drain. When you need to do a water change or if you want to drain the pond, you only need to unscrew the standpipe, opening the drain.

All drains and overflows that exit your koi pond should be screened, even if your smallest fish is considerably larger than the largest-diameter exit. When flowing, a drain exerts powerful suction, and a fish sucked in by its tail will experience serious injury; if it gets pulled into the drain headfirst, it is likely to be killed.

More Pond Topics

TWO FILTRATION OPTIONS

There are two methods of maintaining water purity that rely on natural phenomena and mimic natural settings. The first requires a specific situation for your pond, while the other can be implemented for any setup.

Flow-Through "Filtration"

Aquarists tend to be an impractical lot, content to raise tropical fish in the Far

Marginal plantings soften the look of the pond's edge and form a continuum between land and water.

North, soft-water fish in hard-water areas, and the like. Commercial hatcheries, on the other hand, tend to be much more practical, locating tropical fish farms in Florida, for example. Commercial breeders of rheophilic native species (fish from cool, fast-moving waters) usually are situated where there is a source of cool water that can be channeled to flow through the fish pens and tanks on a constant basis. Many producers of marine species also use a flow-through system—giant clam farmers, for example.

While technically not a filtration system since the water never returns to the setup, a flow-through system eliminates the need for any other filtration. As long as the water source is appropriate for the species being kept, keeping it flowing through the pond is all that is needed, since all wastes are continuously washed down the drain.

If you are fortunate enough to have a suitable spring, stream, or artesian well in or near your garden, you can have a flow-through setup for your koi pond, and all your filtration concerns are over. All that is required is an inlet and an outlet on your pond, with the inlet connected directly to your water source and the outlet connected to a drain.

The simplest version of this is to pipe (or direct in a canal) water from a permanent stream into your pond, then return the water a little ways downstream. The most complex, however, isn't much more involved—the output of an artesian well or spring is directed to the pond, and the

Certain irises are frequently used as marginal plants for ponds as they perform very nicely in soggy conditions.

overflow is routed to the drain. The only requirement other than the water source is that your pond has enough grade. In the event that it doesn't, you can still have a flow-through system, only then you would also need a powerful water pump to move the source water into the pond and a means of piping the overflow to some type of drain—a drainage ditch or dry well.

The advantages of a flow-through system are obvious, but does it have any disadvantages? Certainly. The biggest is that

The volume of water exiting at the lower end of this waterfall provides great aeration, but the height to which the water must be raised from the pool to the top of the falls requires a powerful pump.

your pond is absolutely dependent on the water supply. If it should ever fail or be cut off, you will have no filtration system, and your fish will not last long. Also, if the water source becomes tainted or polluted, your fish will die.

If you are using a natural stream, make sure you have control of it upstream. Your organic farmer neighbor could sell the land to someone who will bulldoze over the stream, diverting it somewhere else and drying up your water supply. The new owner might have a heavy hand with fertilizers and pesticides, rendering the water toxic to your fish. With a spring or well, you must be certain that it is reliable, even in a dry year, and that it is not likely to be contaminated by agricultural runoff. In some areas laws may require that runoff from a pond with fish be chemically treated before being returned to a natural stream, and of course state laws may ban even the accidental release of a non-native fish (such as koi) into natural waters.

Swamp Filters

The use of plants as filters is accentuated in a setup with a swamp filter. Some pond owners use a swamp filter as a supplement to their normal filtration, while others find that it is all the filter they need. Even some stubborn cases of green water have been cleared up with swamp filters.

What is a swamp filter? It is essentially a contained bog through which the pond water is percolated. The plants and the bacteria in the roots and soil process and remove waste products so the water leaving the filter is nutrient-poor, which means no wastes to poison the fish and starvation for algae.

A bog full of plants and bacteria? That may not sound too appealing, but, in fact, with a little planning a swamp filter can be a beautiful addition to your garden.

Depending on the layout of your garden site, you may be able to incorporate the marginal plantings into the swamp filter. Perhaps you can locate the bog some distance from the pond and connect them via a stream or other water course. There should be a grade between the pond and the filter so that water will flow downhill from one to the other. Obviously, you will have to pump water back uphill to complete the loop.

It does not matter whether the water flows from the swamp filter to the pond and is pumped back to the filter or whether it is pumped from the filter to the pond and flows back to the bog by gravity. When they are located side by side, it is possible to have the overflow from the swamp filter just barely above the level of the pond, producing a very gentle cascade of water back into the pond.

How do you make this filter? This is one of those projects where you put in a lot of labor (like digging a pond), but when you're done, there's little to show for your effort (unlike a pond). Despite the lack of visible result, that labor is not in vain. It consists of digging out the area of the swamp filter to a depth of at least 6 inches up to a foot—deeper is better. The area is then lined with a pond liner, and the hole is refilled with soil. If what was dug out is heavy clay rather than sandy loam, it would be best to use only some of it and mix in plenty of peat moss and sand. Although I have never seen it, it should be possible to use just gravel, in which case your bog filter will in fact be a giant hydroponics system, with the waste-laden water from the pond being the nutrient solution.

You need to construct a sump from which the water can overflow or be pumped back to the pond. The easiest way to do this is to place an area of gravel at the outlet end of the swamp filter

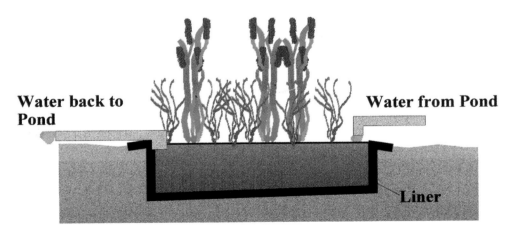

Swamp Filter

Water back to Pond

Water from Pond

Liner

The swamp filter is one way of dealing with summer algae problems.

The water percolates through the soil, passes into the sump, and overflows back into the pond.

Although simply having the water flow through all that soil would produce a lot of filtration, especially once bacteria became established, the plants in a swamp filter are a very important component. The goal is to have every bit of the bog filled with marginal plants that will remove close to all the nutrients and wastes from the water.

The choice of plant species for the filter will depend on your preferences and your climate. There are many beautiful foliage and flowering bog plants that can be used. If the filter is alongside the pond, you can merge the marginals on the shores of the pond with those in the filter, blending the two structures together. This bog is the perfect place for those marginals that usually need to be confined in baskets, since their rampant growth habits are ideal for the swamp filter. If the bog is to be a border or background for the pond or for some other part of the garden, taller plants like papyrus, cattails, and reeds can be placed in the back, while if it is to be viewed from all sides, they should be planted in the center, with lower-growing plants around them.

The only maintenance a swamp filter requires, other than an occasional thinning or pruning of the vegetation, is a periodic inspection to make sure that the water is flowing uniformly through the bog. The formation of rivulets or channels is bad, both because much of the filter is

This three-tier cascading fountain very effectively aerates the small formal pool in which it is situated.

and have the water exit from the graveled section. This keeps the soil from being washed away into the pond. A similar area on the inlet side helps prevent channeling and erosion. Plastic fencing mesh is the best material for the dividers between the graveled areas and the soil.

A less labor-intensive alternative is to create a raised bog. Landscape timbers or other structural supports are put into place, outlining the bog area, which is then fitted with the liner. The bog is filled with a quality garden loam. In this type of setup, a submersible pump can be used to lift the pond water into the swamp filter.

THE GUIDE TO OWNING KOI

being bypassed and because it enables anaerobic zones to develop where there is no constant supply of oxygenated water; the bacteria that thrive in these unoxygenated waters produce toxic substances like hydrogen sulfide.

LITTLE EXTRAS

There are many features you can add to customize and accent your koi pond. Depending on the type of garden (formal or informal), one or more of these options can put the finishing touches on your water garden.

Fountains

Fountains can be excruciatingly classic and formal with several tiers supported by carved water nymphs and Aphrodite at the top pouring an endless stream of water out of an earthen jug, or they can be as simple as a pump return directed upward in the middle of a large rock. Various nozzles are available to change this simple vertical discharge of water into different shapes and spray patterns, and you can add colored lights for beautiful night displays.

Aside from adding to the unique beauty of your garden, fountains perform the beneficial function of greatly improving gas exchange or aeration of the water. The only drawback other than the electricity to run the fountain pump is that all this spraying will also greatly increase evaporation. During hot summer weather this can be considerable, and you will have to be diligent in replacing the water on a regular—probably daily—basis.

Islands

In very large ponds, the central focus

Oriental sculptures and islands can help create a mood in the koi pond.

Children love water gardens. Adults need to supervise their appreciation.

can move from fountain to island. Even in a smaller pond, a very large planted container can be positioned out in the middle for an instant "island," but in big ponds the possibilities are much greater.

If you are constructing a new pond, it is a simple matter to build-in one or more island areas. The pond will basically be a doughnut shape, so if using a liner you will need to account for four times the depth, rather than two, added to both dimensions. The liner that will be cut away from the island land area is waste, but you still need to account for the extra island shores. If you choose this route, drainage on the island becomes a problem. Since it is, in effect, a bag full of dirt, there is no way for water to drain out. If you perforate the liner, water from the pond will infuse the soil up to the water level. There are two ways of dealing with this.

First, you can simply plant the island with bog plants that enjoy having their roots submerged. The alternative is to raise the land level of the island above the water level to a point where you can successfully grow other plants in it. Capillary action will keep the soil moist, but not submerged, so any plant that does not require dry conditions will thrive, since it will always be "watered."

Bridges

Ornamental bridges can enhance your koi pond, and, if large enough, they can serve the practical function of providing shortcuts through your garden. You do not, however, have to have a large pond to enjoy a bridge. In fact, a very effective design is to place a bridge at a narrow end of a pond. It does not have to "go" anywhere—the "other" side of the bridge can even abut a garden wall. Such a bridge will add to the visual interest of the garden as well as provide a spot for quiet contemplation of the swirling colors of the koi swimming under the bridge.

Oriental Lanterns

Japanese-style lanterns, often in stone, are popular garden ornaments that are especially apropos in a water garden featuring koi. These and other statuary can be strategically placed in and around a koi pond to enhance the beauty of the garden. They are often used simply as ornaments, but they can also be illuminated at night, either electrically or with votive-style candles.

Pond Lighting

The presence of a koi pond in your garden offers additional options for using electric lighting to add to the overall effect. Aside from the other types of lighting available for gardens, special underwater lighting can extend your enjoyment of the koi pond into the night hours. Obviously you should use only approved devices that are certified safe for underwater use, which brings up the whole idea of pond safety.

POND SAFETY

This section is short and to the point. General commonsense safety guidelines should direct your use of the pond, such as keeping children under constant supervision and designing paths and walkways to minimize the threat of people stumbling into the pond. The presence of water, however, presents a whole other area of concern—electrical safety.

Ponds can require many electrical devices—pumps, filters, UV sterilizers, and ornamental lights. To protect yourself and your family, you must make sure that all such devices are manufactured, installed, and maintained to ensure safe operation. Have a certified electrician install all the equipment according to national and local codes. Primary to the safe use of electricity outside and around water is the use of GFI (ground fault interrupting) circuitry. Put simply, if you do not protect all koi garden electrical objects with GFI circuits, you are playing with the lives of the people who use the garden.

Plantings

Koi can live in a pond devoid of all plant life and situated in a plot of macadam, but the use of aquatic plants not only shows off the fish nicely, it improves the conditions in the pond. Your koi will be set off even more nicely by adding various plantings around the pond. When the pond is totally integrated into the garden so that it is impossible to say exactly where one ends and the other

Koi can live in an unadorned pond, but most people prefer a full complement with fish, plants, and decorative features.

THE GUIDE TO OWNING KOI

The vibrant freshness of this hardy yellow water lily contrasts nicely with the crisp green of the lily pads.

begins, the effect is much more natural and beautiful than simply having a pond surrounded by a garden. This can be accomplished in many ways, but the easiest is to use a stream or water course.

THE STREAM

It is a simple matter to construct a stream using a liner. Liners are sold in rolls 5 or 6-feet wide for this purpose. Your imagination is the only limit, but a common and effective design is to pump the output from your filter system uphill and into a rock waterfall, then to let the water run downhill through the stream back to the pond. Not only does this provide excellent aeration, it gives you a gurgling stream to landscape.

The artificial nature of such a stream is easily hidden by several means. Dig a meandering rather than straight course for it and use overlapping and overhanging flat stones to cover the edge of the liner. Place gravel along the bottom of the stream and put stones or boulders of various sizes at strategic places to create riffles, cascades, and "rapids." Then you can plant the banks of the stream with marginals and merge them into rock gardens or perennial beds.

THE WATERFALL

A garden waterfall adds a special touch, and it can range from a pump outlet at the top of a rock pile to an elaborate series of cascades and pools, with ferns and flowering marginals rimming each. Besides providing you with a wonderful addition to your garden, the waterfall—like the stream—adds beneficial aeration. If your

Regardless of how many other plants are used, water lilies almost always form the central point of interest.

garden has a natural slope, the waterfall can be integrated into the regular plantings to effectively bring them together with the water features. But what about the aquatic plantings?

SUBMERGED VEGETATION

We've already discussed the role that the so-called oxygenating plants have in algae control, especially until tougher plants like lilies can become established. While it is not difficult to get these plants growing while your koi are small, they can be difficult to maintain in the presence of larger fish. Some koi owners simply forego having them in their ponds at that point. Others value them so highly that they provide a fenced-off area of the pond where they can grow unmolested and still contribute to the ecosystem of the pond. It also is possible to use half-inch plastic mesh to build cages around these plants; as they grow through the mesh, they are grazed by the koi. Mesh around the plants is not a bad idea even for tough plants that the koi will not eat (like water lilies), since it keeps the fish from digging around their roots.

WATER LILIES

A koi pond without lilies, a lily pond without koi—both seem somewhat lacking. The synergism of koi and lily is not just esthetic, either. Lilies are heavy feeders, so they benefit from the nutrients that koi produce in their wastes. They should be protected from the natural rooting and digging of the koi, however. A plastic mesh barrier can be used, or you can cover the surface of the lily pots with large gravel to keep the koi away from the soil.

Lilies are prolific bloomers and will produce flowers throughout the normal pond season, year after year. There are three types of lilies for your koi pond—hardy, tropical, and tropical night-blooming.

Hardy Lilies

As their name states, these plants can be overwintered in most ponds. They should be potted in fertile soil, with the pots positioned on shelves or supports to locate them about 18 inches below the surface. There are several species of hardy lilies available in basic colors of whites, reds, and yellows.

If the roots (tubers) of the lilies in your pond are likely to be frozen solid, you will need to protect them during the winter. The easiest way to do this is to wait until the plants have died back in the fall, then move their pots to deeper water, where they and the koi can enjoy frost-free conditions all winter. Just remember to move them back to the proper depth in the spring!

Tropicals

Tropical lilies need more protection than hardy lilies, but many northern aquatic gardeners find the extra work worth it. The beauty and variety of tropical lilies parallel the hardy ones in the whites, reds, and yellows, but they add blues and purples as well. These plants typically prefer less water depth, with most requiring planting between 6 and 12 inches deep. Temperatures much below 60ºF are fatal to these lilies, so after they go dormant in early fall you must either move them to a greenhouse location or store the dormant root in a cool, moist place until spring.

Night-blooming Lilies

It is among these delicate tropical species that we find the exotic night-blooming water lilies. Their care is like that of other tropicals, but they add their

An aggregation of *Nymphaea* "Conquerer," a hardy red variety. The bloom of this variety and most others keeps one color permanently, but in some varieties the color of the bloom changes from time to time.

The *Nymphaea* cultivar "Jennifer Rebecca," a night blooming variety.

The marsh marigold is found in almost every water garden. It is extremely hardy and has bright yellow flowers early in the season.

floral contribution to your pond's nocturnal beauty. It is hard to beat the sensuous effect of a summer night's moonlight on a garden pond accented by these beautiful blooms and the occasional splash of a glistening koi at the surface—and the koi even keep down the mosquito population!

MARGINALS

I've never liked this name, since it implies peripherality, and marginals can, in fact, be a major attraction in your garden. The name refers, of course, to the fact that these plants are naturally found at the margin between water and land—in bogs and shallow water areas. In your garden, however, you can use these bog plants almost anywhere—naturally along the edge of the pond and in and along the stream and waterfall, in artificial "wetlands" created with a pond liner, or even in the middle of your pond planted in pots or baskets and raised to the appropriate depth on piles of stones or bricks.

Since this grouping of plants is made simply on the basis of their preference for growing with wet feet (in 0 to 6 inches of water), it necessarily includes a wide range of plant types. Some of these plants are grown for their foliage alone, but many flower profusely as well. An enormous variety is available, from large cattails (*Typha*) and towering bullrushes to flowering compact species like the marsh marigold (*Caltha palustris*) and aromatic foliage plants like the aquatic mint, *Mentha aquatica*. Some very interesting stepped effects can be made by using dwarf cultivars of taller plants such as a stand of cattails behind a row of dwarf cattails or large papyrus (*Cyperus*) in back of dwarf types.

The Koi

If you start looking into the varieties of koi, you might get the impression that

This is a grand champion koi of the kohaku type. The body has imposing proportions with bright red coloring equally distributed over a white background. There is a sheen to the scales that can only be achieved with maturity.

you'd have to learn to speak Japanese and maybe even go to Japan in order to get very far. The Japanese predominance in this field is evident in such things as the official names of the types of koi all being in Japanese, but the koi hobby is alive and well in the United States, too. Even if you are interested in show stock, there are breeders, shows, and auctions all across the country where you can get the highest quality koi—in some cases imported from Japan, in others, domestic-bred.

Many beautiful koi do not conform to the strict standards of the official color types and are available at much more reasonable prices to the pet trade. You will be able to get nice fish from aquarium stores, water garden suppliers, and commercial fish hatcheries. On the other hand, you may choose to start with show stock from an experienced koi breeder. In any case, the care of a five-dollar fish and a five-thousand-dollar fish is the same, at least in theory—I imagine owners of the latter are a tad more careful in the husbandry of their fish.

A JAPANESE LESSON—THE VARIETIES

If the correct Japanese terms for the varieties of koi do not interest you, simply read this section for the descriptions and photos of the different color types and ignore the names. In any event, you can use this chapter as a reference when locating and ordering your koi so that you

The scarlet circles of this taisho sanke start at the head and repeat on the pure white background. The shiny black encroaching onto the white adds a special touch, and the white tail gives a precise finish.

This ginrin showa sanke is dazzling beauty. Round, shiny scales run along its backbone, and with these as a base, the strength of the round red markings moves into a red design. Lustrous black punches through with boldness and beauty from the left shoulder moving to the right mid-portion and continuing on to the left in perfect balance.

This shiro utsurimono has a fine physique. The black markings starting at the mouth are sharp and clear as they break into its white texture. This is high quality black. The whiteness that suddenly opens at the shoulders leaves a strong impression and the dorsal fin is exquisitely beautiful.

make sure to get the varieties you want.

Remember that koi are supposed to be viewed from above, not through a glass side panel. The centuries of breeding that have produced the modern koi varieties were always focused on looking down on the fish, and you will almost always see koi photographs with this orientation.

Kohaku

The kohaku is the oldest variety and still the most popular. It is a white fish with red spots. The number of spots can be reflected in the name of the variety:

nidan kohaku—two red spots;

sandan kohaku—three red spots;

yodan kohaku—four red spots;

godan kohaku—five red spots.

When I say "red" here, I mean a brilliant blood-red. This is not a faded orange or pink. The stark contrast of the deep red on the white background makes for a gorgeous fish. The red splotches are typically located down the spine.

Sanke

The sanke is a tricolored fish. It can have a background of white or of black, with splotches of red plus black or white. The two varieties are:

tashio sanke—a white fish with red and black markings;

showa sanke—red and white markings against a black ground.

Utsurimono

This name means "reflections," and these are black fish with markings of a single color. There are three types:

shiro utsuri—white markings;

hi utsuri—red markings;

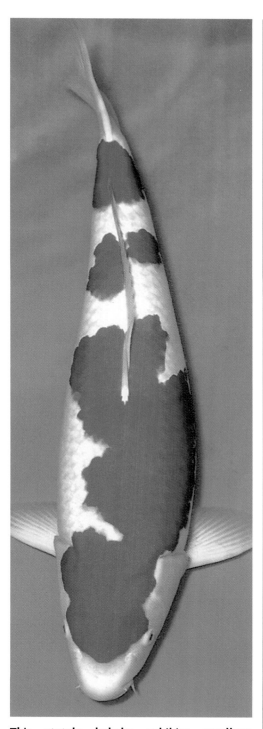

This stately kohaku exhibits excellent uniformity, depth of its red markings, and fine edges. The markings are extremely strong, and if it grows to be a jumbo koi, it will be even more exciting to observe.

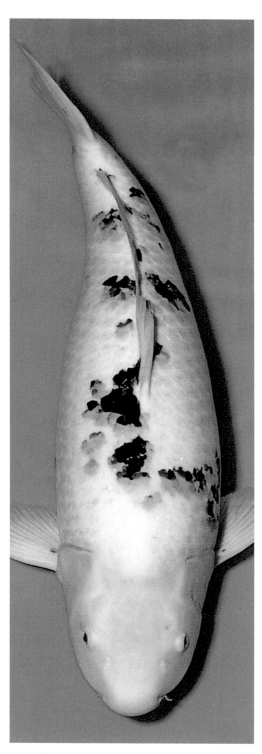

Pure white skin on a magnificent body and fine quality black markings make this shiro bekko a splendid choice.

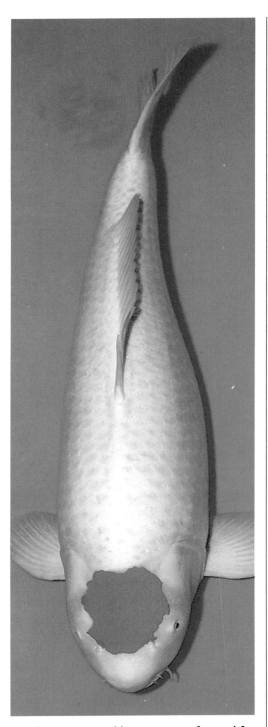

Tancho are rare, and large ones are few and far between. This koi is blessed with a fine physique. The superior quality red marking on a deep white background raises it to an even higher level.

The bright red is seen rising up out of an indigo background, so typical of goshiki. This goshiki displays a fine balance between white and indigo.

Here is a koromo with a magnificent physique. The fine quality of the red markings has been developed to the fullest.

The two well-balanced and fine quality black markings on the shoulder and back are the strongest points of this doitsu.

As koi become larger, they have a tendency to show defects and wrinkles, but there is not a single flaw in the gold color of this utsurimuchi.

ki utsuri—yellow markings.

These are startlingly beautiful fish with bright spots on the black ground.

Bekko

The bekko is somewhat the reverse of the utsuri. It has a white, red, or yellow background with black splotches running down the spine. There is no black on the head. The varieties are:

aka bekko—red fish with black markings;

shiro bekko—white fish with black markings;

ki bekko—yellow fish with black markings.

Asagi

The asagi has a very light blue head, darker blue reticulated scales on the body, and red on the belly, the bases of the fins, and the gill plates, for a striking combination of colors.

Tancho

This is a fish with a circular red spot on top of the head. It is not a breed in itself, but it is a very beautiful variety. Variations are known as:

tancho kohaku—overall white with the circular spot as the only red;

tancho sanke—white with black markings and the only red being the circular spot on the head;

tancho showa—black with white markings and the only red being the circular spot on the head.

The tancho kohaku is the most popular, since it is colored both like the *tancho*, a white crane with red over the top of the head that is the Japanese sacred crane, and like the Japanese flag—a red circle on a white ground. The tancho is not a fixed

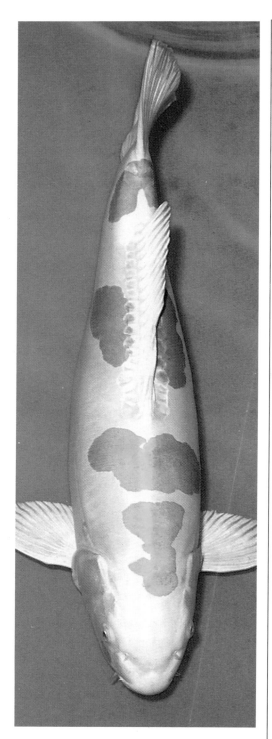

breed but just the chance concentration of all the red coloration on top of the head. Breeding two tanchos produces all types of kohakus and often no tanchos.

Goshiki

This is a blending of the sanke and asagi patterns—a white fish with black and red markings plus the reticulated blue. The tricolored pattern on the white ground makes an interesting "calico."

Koromo

Another blending of sanke and asagi, this time with the reticulations being restricted to the red spots. This makes a very interesting pattern as well.

Doitsu

The Japanese rendering of Deutsche (German), this trait was bred into koi from the scaleless German carp. These fish either are without scales or have lines of large scales down the middle of the sides and down the spine. There is a doistu version of almost every color variety.

Hikarimuji

These are solid-colored fish with a metallic sheen. There are several varieties:

yamabuki—solid metallic yellow;
purachina—solid platinum;
orenji—solid metallic orange;
nezu—solid metallic gray.

These are also called *ogon*. Related to the hikarimuji are the *matsuba*, which have a pinecone pattern of black over a solid color. The black is restricted to the center of each scale, producing the pinecone effect.

Hikariutsuri

These are showa sanke or utsurimono

Hikarimoyo is all "look at me!" brightness with the head reflecting the most light from a pattern that is arranged in a well-balanced manner from the shoulder to the tail.

The pattern is extremely simple, with deep silver on a pure white skin. The red is soft, and though in small designs, the black markings are of high quality, and the body shape is ideal on this ginrin sanshoku.

This kumonryu has a very strong and powerful body. The black of the body is of fine quality and balance.

This is a jumbo kujaku. The red pattern on a shining background makes a truly elegant statement. The scale pattern is highly desirable.

A modern showa type with unrivaled natural features. Beginning with the soft lines on both sides of the head, the distribution and placement of color makes a marvelous impression. The deeply lustrous black on the bright upper part of the body is tremendously appealing.

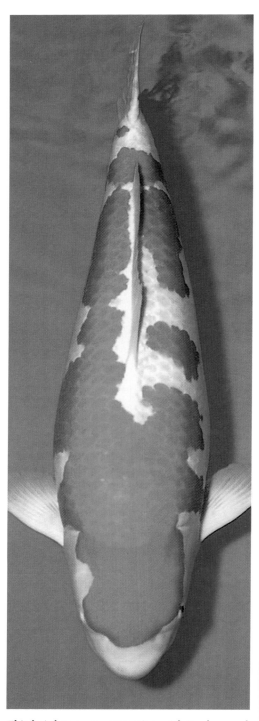

This koi draws your attention with its deep red markings, a grounded white background, and impressive physical presence. This is a kohaku whose natural attributes have been fostered by finely tuned rearing skills.

crossed with hikarimuji, creating those varieties with the metallic sheen. Three varieties have been developed:

kin showa—showa with metallic sheen;

gin shiro—shiro utsuri with metallic sheen;

kinki utsuri—ki utsuri with metallic sheen.

Hikarimoyo

All other metallic patterns are called hikarimoyo. Varieties include:

sakura ogon—metallic kohaku;

yamato nishiki—metallic sanke.

Kujaku

This is a blending of the metallic trait with a goshiki pattern. Its full name, *kujaku ogon*, translates as "peacock."

Kinginrin (Ginrin)

This is another metallic trait, where the scales reflect light like little mirrors, giving such varieties as ginrin kohaku, ginrin showa, ginrin sanke, etc.

Kumonryu

This breed, which translates as "nine-crested dragon," is a doitsu-scaled fish that is white with black markings. The markings, however, change over time, and a fish can go from all white to all black, with any variation in between. Various factors including temperature and maturity are thought to play a role in the pattern changes. The kumonryu has only been recognized since 1994.

JUST GETTING STARTED

This doesn't come close to exhausting the more than one hundred classifications for fixed strains of koi, and there are many

This is truly a powerful taisho sanke. This magnificent koi has red and black markings in fine balance on a white background.

varieties that have not yet been named and standardized. It does, however, cover the basic types of koi you are likely to encounter. If you just want beautiful fish for your garden pond, the names, of course, are not important, and many very beautiful koi fit under various "all other variety" categories anyway. The only real consideration for what color koi to get for a garden pond, other than your personal preferences, is price. The rarer and harder-to-breed varieties naturally cost considerably more than the more common ones, like the kohaku, and well-patterned show fish are much more expensive than non-show-quality animals, which can nevertheless still be very beautiful.

Since many of the patterns develop slowly as the fish mature, you often can purchase juvenile koi of uncertain quality at considerable savings and raise them yourself. Not only is this enjoyable, you can get a nice variety of colors and patterns this way. It is unlikely that any of the fish will be ugly—although show koi breeders might turn up their noses at some of them.

Feeding Koi

Feeding your koi need not be complicated, but it must be based on a firm understanding of their nutritional needs. These needs vary depending on the maturity of the fish and the temperature of the water.

COUNT THE RINGS

The age of a koi can be determined by examining one of its scales. The scales (and the rest of the fish!) grow more quickly in warm weather, and in very cold weather they do not grow at all. This produces definite bands of growth or "rings" on the scales. You can tell how old a koi is by the number of rings on its scales, the same as you can with a tree by counting the rings in a cross-section of its trunk.

The significance of this for feeding is that a koi's need for nutrients changes with the calendar. In the spring, when the water warms, the fish begin active feeding. They need large amounts of high-protein feed to fuel their rapid growth during this time. Koi can grow as much as a foot or more their first year. Once they are sexually mature their growth slows down considerably, but they will continue to grow, though almost all of the growth will be during warm weather. As the weather cools in the fall, they need less food, and they cannot handle high-protein foods any more. It is important to feed grain-based feeds that are easily and quickly digested. If foods with a high level of fish-meal or other proteins are fed at temperatures below 50°F, this difficult-to-digest food may stay in the gut long enough to actually putrefy, which will prove fatal.

THE TEMPERATURE CONNECTION

Knowing your pond's water temperature, then, is important before choosing your fish's feed. As stated, below 50° only a low-protein diet can be tolerated. Many proprietary feeds, often based on wheat germ, are formulated for cool-weather

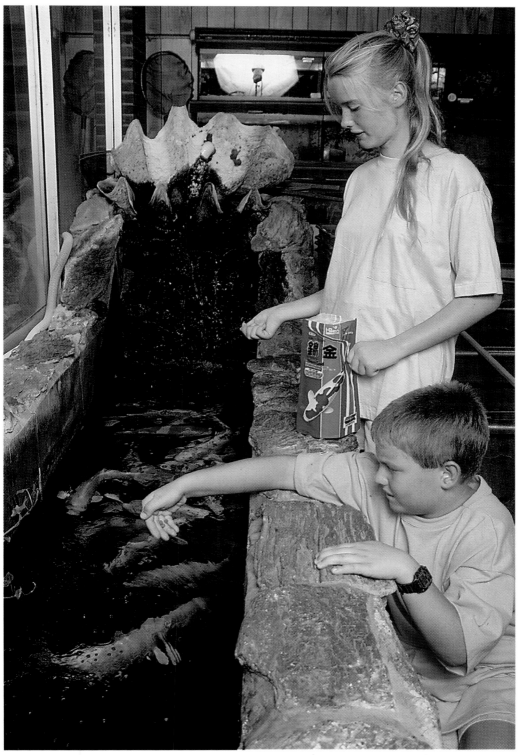

Koi are wonderful beggars. They are all open mouth and "feed me" attitude. Be careful not to overdo it.

feeding. As the temperature continues to drop, the fish will stop feeding completely.

Above 50° you can begin to feed regular koi foods, and by the time the water reaches 60°, you will need to use a high-protein growth diet. Such feeds are formulated with as much as 40 percent protein. At very high summer temperatures your fish's appetite may decrease, but they can still handle the high-protein food.

Of course, if you keep your koi indoors in the winter, if you heat your pond, or if you live in an area where the water temperature will never drop below 50º, then you will be able to feed growth foods for a greater part of the year or even all year, and your koi will grow that much faster.

PELLETS OR STICKS?

Commercial koi food is available as pellets or extruded "sticks" and in floating or sinking formulations. Obviously, the size of the pellets or sticks will depend on the size of your koi, but you must remember to consider the smallest fish in your pond when choosing the food. Remember that if you mix large and small pellets, there is nothing to stop the larger fish from consuming all the small ones then leisurely munching on the big ones while the small fish peck at them futilely.

Koi, being carp, naturally feed on the bottom, so sinking formulations are a perfect choice. On the other hand, koi, being carp, are eager, greedy feeders and will quickly learn to come to the surface for dinner, so if you enjoy watching your fish

eat, floating formulations are also a perfect choice. Floating foods additionally help you to gauge how much to feed, since it is easy to see if everyone is getting enough and if there are leftovers at every meal.

It is always a good idea to feed a variety of foods, since even fish get bored with the same thing to eat every day. This also gives you an opportunity to round out your fish's diet, since if you feed several different foods, it is likely that any possible deficiencies in one food item will be offset by one of the others. Many koi owners enjoy interacting with their fish, which are very intelligent. A diverse diet can also include treats and extras that are very useful for taming and training koi— they will even learn to take food from your hand, actually poking their heads out of the water to grab it from you.

OTHER FOODS

Besides a variety of commercial cool- and warm-weather diets, your koi will appreciate treats of live and fresh foods. One reason koi in an outdoor pond are so much more colorful is the natural food supplements on which they can feed. Insects that fall into the pond, plus insects and other invertebrates that live in the pond itself, provide important pigments that enhance the fish's colors. Algae also are an important stimulus for the brightest colors, and there is typically no shortage of this in a koi pond.

You can supplement your koi's diet with earthworms and spirulina-based foods,

The benefit to hand-feeding your fish is that you can be sure the food is actually being eaten and not falling to the bottom to rot.

which will give them a treat, improve their nutrition, and help them achieve maximum growth and color. Fresh, dark green leafy vegetables such as romaine lettuce and cereal foods such as cooked oatmeal and whole-grain bread are enjoyed by koi as well.

PREDATION

Now that we've discussed what koi like to eat, it is time to talk about things that like to eat koi. Unfortunately, fish kept in outdoor ponds are at the mercy of many predatory animals.

Birds

Various cranes, herons, egrets, and other birds are eager fish-eaters. Depending on where you live, these may or may not be a major consideration. Even where these birds are common, they tend not to visit ponds close to human habitations, so many gardens are relatively safe from their predation. It is difficult to pro-tect your koi against these birds, but most of them will only take juvenile fish and are more of a concern in breeding ponds. Having the pond slope rapidly away from the shore is a great deterrent, since these birds fish while wading, and this way the fish can escape to deep water quickly.

One concern about birds like this visiting your pond is that they can bring parasites and diseases to your fish. Even large invertebrates like snails and leeches can hitch a ride under the claws of a heron and move from pond to pond. Ducks, which may visit your pond, can also transport pathogens. It is unlikely that songbirds sneaking a quick bath or a drink at the water's edge are reason for concern, but it is best to discourage any waterfowl, whether predatory or not, from visiting.

Masked Bandits

Although children's books and animated movies make it out to be a cute, cuddly, friendly creature, *Procyon lotor*, the raccoon, is a vicious, ruthless, wasteful killer. I have had them reach their fingers through the half-inch by inch mesh of cages and pull baby turkeys through piece by piece. Once one broke into my chicken coop and skinned the head off a chicken, leaving the bird bleeding, blind, and wandering aimlessly in circles—this after already killing several of the birds in the coop and eating at most one bite out of each.

More importantly, they love fish. They also love wading in water, feeling around for squirming goodies. They also are ubiquitous, so you cannot avoid them—they actually may have higher population den-

This ornamental heron may deter others of the feathered kind from raiding your fish stock.

sities in many urban areas than in wilderness regions.

Well, if they're that prevalent, what can you do about them? That's the problem. There is no single sure-fire solution to the raccoon problem. Because they are highly intelligent animals, they learn quickly, and what works in one situation at one time may not work in another. Electric fencing, if properly installed, is usually effective, but these varmints are experts at finding chinks in the armor. Depending on your garden layout, you may be able to install several wires only a couple of inches above the ground and spaced a couple of inches apart. These are easily hidden by low-growing plants. Short sections of PVC pipe, drilled to receive the wires, make excellent insulating fence posts for such a setup. By making it necessary for the animals to traverse several hot wires, you almost guarantee that they will be zapped before reaching the pond. In most cases that will be sufficient, though they are sure to discover rather quickly if your fence charger malfunctions—and it wouldn't surprise me to learn of a raccoon that learned to parachute drop inside the electrified area. You only need to turn the charger on at night, since daytime raccoon predations are rare. The same setup can, however, protect against daylight visits to your pond by cats and dogs.

Probably a deep pond is the best all-around insurance against coons. Although they are powerful swimmers, they hunt while wading, not diving, and as long as your fish can retreat into the depths, you

These bandits are tough to deter.

are not likely to lose many to these deceptively cute but extremely malicious animals.

Other Mammals

In most garden settings other mammals like dogs, cats, skunks, and opossums may pose some threat, though deep water areas are usually sufficient protection against them as well. If aquatic or semi-aquatic mustelids like otters or martens inhabit your area, there may be no way to protect your koi other than to complete-ly cover the pond with wire mesh fencing.

Potentially the most dangerous mammalian predator is, of course, *Homo sapiens* (man), and owners of valuable koi often have to resort to elaborate alarm systems, guard dogs, and other security devices to protect their investment. Your garden pond of koi is much less likely to be a target of knowledgeable thieves than

a breeding operation would be, but you should be alert to the possibility of van-dalism. Floodlights operated by motion detectors are effective deterrents, and they offer some protection against four-footed vandals as well.

Reptiles

A garden koi pond is not usually the type of habitat that supports water snakes or snapping turtles, the worst offenders among reptile predators of fish. In any case, they are only a danger to small koi. If, however, you live in or near Florida, an area populated by alligators and (in the extreme South) crocodiles, one could show up in your garden pond and have a feeding frenzy, though in this case the loss of your koi would only be one of your problems. A low fence is nor-mally all you need to exclude these ani-mals from your garden, however.

Breeding Your Koi

Very few people who have a koi pond breed their own fish. Your koi will likely spawn, but the eggs and fry will be eaten. A few may survive, especially in a heavily planted pond, but intentional breeding usually is not the goal of garden koi pond owners. Oh, many koi breeders will have a decorative pond or two for displaying them, but koi breeding is an involved undertaking requiring a commitment well beyond that of the typical garden pond owner, and it is best undertaken in ponds that leave a lot to be desired if considered only as ornaments.

SELECTION OF THE FINEST

Whenever human beings have taken an animal species and selectively bred it for various traits, a cult grows up around this animal—an elite subculture where pedigrees and show ribbons are of utmost importance, and culls are labeled "pet quality" (or in the case of livestock, "meat quality"). Phenomenal sums of money are exchanged for top animals, and their offspring are in highest demand.

While these breeders are very intense, they represent a minority of the people who raise and keep animals, albeit a very important one. The trickle-down effects of their fervor are what maintain the quality of the gene pool, enhancing desirable traits and developing new ones. It is often the case that "faults" that disqualify an animal from the show bench are of little significance anywhere else. The pet or agricultural value of an animal is not affected much by these minor deviations from an arbitrary and idealized standard. This is because the overall set of characteristics for the breed makes high-bred animals also well-bred animals.

The major traits that make a Charolais bull a fine show animal and stud prospect also make it a great beef animal. Dog

Healthy adult koi will spawn in the spring.

breeders who ignore temperament and other important pet qualities will quickly lose their market. This means that the elite show set tends to produce the best animals, and, since they are very picky, many of the animals they produce will not be up to their high breeding standards and are sold to the general public. These people's investments of time and money bear good fruit, but they typically also make it unprofitable for part-time breeders of these animals. This is not to say that you cannot be successful breeding koi on a small scale. In most cases, however, people who attempt this either wind up quitting after a while or they make the step up to the show circuit.

SPAWNING

Healthy koi will probably spawn each spring. The males will chase the females into the planted shallows and spray thousands of eggs. They will also consume thousands of eggs, as well as any fry that manage to hatch. If there is sufficient vegetation in the pond, a few fry may

escape until they are too big to be swallowed, at which point they will come out of hiding and join the school. If, however, you want to raise some koi intentionally, you will need to intervene.

This is not an undertaking that should be begun haphazardly. A single spawning pair of koi can produce about 50,000 fry. Even if you followed the custom of Japanese breeders and culled half the fry at hatching, you would still wind up with 20,000 or 30,000 juvenile koi. It will cost many hundreds of dollars to feed them, and it will require hour upon hour of water changing, feeding, culling, and separating the juvenile fish. Filtration must be rigorous to cope with the massive wastes but gentle enough not to damage the fry. Some of the coloration patterns do not develop and therefore cannot be judged until the fish are one or even two years old.

Unless you have your own private lake in which to house these fish, you are obviously going to run into trouble long before you can effectively judge their relative merits and worth. Certainly rigorous and repeated culling is necessary, but even so you will have a lot of koi to dispose of.

Think about it. Here is a species that produces 50,000 offspring per spawning, yet prize specimens can command prices of $50,000. That means an awful lot of fish produce very few winners. The practical concerns of divesting yourself of thousands of koi should cause you to consider carefully before starting a koi breeding project.

Before you start calculating... *well, even at only $5 per fish, each breeding pair will gross a quarter million dollars!*...and before you start looking at yacht brochures, you must realize that you will only be able to sell a small per-

centage of the fry. Very little demand will exist in your immediate area, and the costs and hassles of running a shipping concern as well as raising all those fish will almost certainly overwhelm you in a very short time. On the other hand, raising a few fry for your own use can be a very rewarding part of having a koi pond in your garden.

LOW-KEY BREEDING

This low-stress, low-expense, low-hassle breeding method focuses not on raising a spawn, but simply on getting a few offspring from your pets to increase the size of your school and perhaps the variety of types you have.

Professional koi breeders produce thousands of fry and cull carefully to ensure that they are raising only the best of each spawn.

If you have a spare pond or a suitably large vessel to use for spawning, you can select the female and place her in this for conditioning. Feed her heavily with live foods, and she should soon develop a gravid plumpness. Place a large quantity of live plants or artificial spawning medium in the pond and introduce your best male koi. They should spawn soon, after which you should remove them both.

When the fry hatch and become free-swimming, net most of them out and dump them into your koi pond. Few if any will survive, but at least there is a chance. How many you leave will depend on your facilities. You might want to keep two to three times as many fry as you intend to ultimately add to your collection. This will give you enough to cover any losses as they grow. You will also be able to cull all but the finest of the fry, and, since we are talking at most dozens rather than tens of thousands, you will probably find a local market for any fish you decide not to keep.

RAISING THE FRY

One of the best places to raise koi fry is in an algae pond. Green water, the bane of an ornamental koi pond, is ideal for baby koi. Not only do they eat the algae and the organisms that feed on the algae, the green water will support a thriving colony of daphnia and other tiny crustaceans. At first the fry will be unable to eat the adult organisms, but they will feast on the young ones. Very often by the time the daphnia have cleaned up the green water,

Spawning mats help protect the eggs.

the koi have cleaned up the daphnia. This system gets the fish off to a flying start. You should have good filtration and perform regular water changes to maintain the water quality.

In the absence of a greenwater pond, you can substitute a multitude of fry foods, both living and prepared. Because there is not an abundance of algal growth to take up excess nutrients, the water in your rearing pool must be very heavily filtered and changed even more often to prevent a deterioration of conditions.

In warm weather, with lots of good food and plenty of clean water, the koi will grow fast—up to a foot or more the first year. Once they are too large to be eaten by your adult koi, you can select the few you want to keep and try to find homes for the others. Healthy juvenile koi (in small lots) are sure to be of interest to pet shops and aquatic gardening stores, and you will have the satisfaction of adding some home-grown fish to your pond!

Health Issues

Even if you don't normally subscribe to the ounce-of-prevention, pound-of-cure notion, it is easy to see that preventing disease among your koi is going to be a lot easier and less expensive than medicating a couple thousand gallons of water and knocking out your pond's biofilter. The husbandry techniques already outlined in this book will go a long way to preventing health problems in your fish—good filtration, no crowding, nutritious and varied diet, water changes, etc.

LIFE ON KOI

Most common health problems with pond koi involve not the bacteria and viruses we normally think of as associated with illness, but parasites. Many parasitic higher organisms plague fish, taking up residence on their bodies or in their gills and causing disease. Fortunately, since they are found on the outside surfaces of the fish, they are more easily affected by medications and other treatments than they would be if they lived inside the koi's body.

Ich

If you are also an aquarist, you know this old foe. Ich, or white spot disease, is caused by the protozoan *Ichthyophthirius multifiliis*, which appears as a sprinkling of white spots all over the affected fish. It has a free-swimming stage (swarmers) prior to settling on the host fish; this is when the organism is most susceptible to medications and other treatments. It is highly contagious and often appears after a period of stress on the animals. There are proprietary medications that will eliminate ich even in a large pond, but since the causative organism is only vulnerable for part of its life cycle (the swarmers), you will have to repeat the treatment until you kill all generations and prevent reproduction.

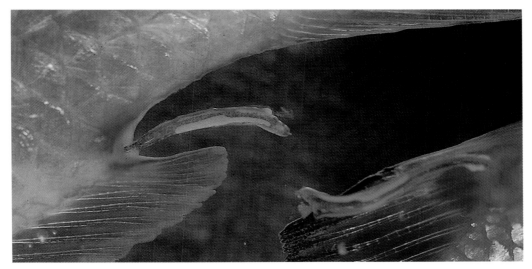
Remove anchor worms and treat the wound with an antibiotic to prevent infection.

Anchor Worms

These parasites are actually copepods (crustaceans) of the genus *Lernaea*. They are easily recognized by the two egg sacs hanging from the body, which attaches to the fish. Not only are the animals themselves harmful, but the wounds they make attaching to the fish can become infected and ulcerated. They should be removed with tweezers, and the wound should be swabbed with a topical antibiotic. There are commercial preparations to treat the pond as well.

Fish Lice

A koi infested with these flattened crustaceans, *Argulus*, will have tiny transparent "bugs" crawling over its body and often will be seen attempting to "scratch an itch." Effective medications are available from your dealer.

Gill Flukes

These trematodes (flatworms related to planarians) are unfortunately common in pond-reared fishes of all kinds, and koi are no exception. An afflicted fish will clearly be having respiratory difficulties—labored breathing and anemic (pink rather than red) gills. A microscopic examination is necessary to identify these parasites, which should be treated with a proprietary medication specifically recommended for flukes.

SIMPLE PATHOGENS

There are, of course, bacterial and viral illnesses in koi. They are not easily diagnosed or treated, and if you suspect their presence, you should seek the assistance of a qualified veterinarian. As you might suspect, the incredibly high monetary value of some koi has stimulated considerable research into koi medicine, and veterinarians who specialize in koi and other commercial fishes are becoming more common. From immunizations to surgery, the treatments available for koi are numerous. Of course, there is a different urgency to treating a koi that costs less than the doctor's visit and one that costs as much as a new car.

THE SALT SOLUTION

I recently read a rather detailed piece on the treatment of koi diseases and was struck by something that I agree with but rarely see mentioned. Although each section on a disease or parasitic affliction contained up-to-date information about the drugs available to treat sick koi, each section also listed salt as a treatment. Salt, good old sodium chloride. It was listed as a treatment or preventative for virtually every ailment, and it received separate attention in a section on the synergistic effect salt has with many medications, increasing their value.

One of the nicest things about salt is its low price. When you figure the dosage for a pond containing several thousand gallons of water, you will quickly realize why salt has remained the treatment of choice for many pond owners. Almost any type of salt can be used, provided it does not have any chemical additives. It is important that you read labels carefully. For example, some "solar salt" (which is produced by evaporating sea water) is pure and additive-free, but other brands include various anti-lumping, iron-removing, and resin-cleaning substances that can harm or kill your fish.

Salt is safe. Koi tolerate it very well, and some people keep some salt in their ponds at all times as a general tonic. You can, of course, use too much and stress or even kill your fish, but the margin for error is extremely wide, and without going to the extreme, you will not overdose your fish.

So does salt have any drawbacks? Well, although it can be used to treat just about everything, it isn't the best or most effective treatment in some cases. It is not a wonder drug or a cure-all miracle; instead, it does directly affect many disease organisms, and when it doesn't, it helps the fish's immune systems by relieving stress, increasing slime coat, and fighting secondary infections. It is also very useful for prophylactic dips. Many ectoparasites and disease organisms are much less tolerant of high salinity than are fish. In fact, koi can easily handle full-strength sea water for a minute or two, but in this length of time many organisms attached to the fish's fins, skin, or gills either die or let go and drop off the fish. This is an excellent way to treat newly acquired koi before settling them into their quarantine quarters.

QUARANTINE

When first setting up an aquarium, we often consider the tank to be its own quarantine tank and only use a second tank for quarantining later acquisitions before adding them to the main tank. With a garden pond, however, there are good reasons for having a quarantine vessel even for your first koi. Primary among them is the fact that inspecting the fish for signs of disease is very difficult in a pond, and it is expensive and inconvenient to treat an entire pond if disease does manifest itself. There are many containers you can use for quarantining your koi when you first get them that make inspecting and treating them much easier.

Children's wading pools (make sure they are not treated with an algicide, which can kill fish), plastic stock tanks, fiberglass bins, large aquaria, and other inert, water-tight vessels all make good quarantine tanks. They are easy to move around, fill, and empty; they make it easy to watch your fish's progress; and they can be completely disinfected between uses. Vigorous filtration and aeration are needed in such tight quarters, and daily or twice-daily water changes are recommended, as they are in most quarantine situations.

What do you watch for? Signs of disease or infection include red splotches, cottony growths, white spots, black spots, ulcers, gasping at the surface, labored breathing, listlessness, unusual stools, and obvious parasites. If nothing shows up in two or three weeks, you can release the koi into their permanent home. It is extremely important that you quarantine any additional koi for two or three weeks *minimum* before adding them to your established pond.

HANDLING KOI

Netting is not a good way to capture and move koi. During the chase and capture they can be injured, and their slime coat can be scraped off, leaving them open to infections. The ideal way to move a large koi is to place the transport container underwater and "herd" the fish into it. By quickly tipping it up, you can get the fish without having to touch it with anything.

The next best alternative is to catch and move the fish by hand. A wet human hand

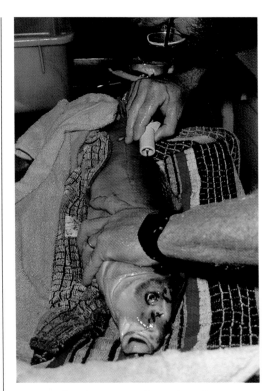

Protect the slime coat of your koi. If you must handle the fish, hold it in a wet towel.

can be much gentler on a koi than a net. Be careful to support the fish with both hands and get it back into water as quickly as possible.

If you must net a koi, use two nets—a chase net and a capture net. Drive the fish with the one into the other, and don't chase it around with a single net. This will exhaust the fish (and probably you, too) and will greatly increase the chance it will injure itself by bumping into solid objects in a frenzy to escape the chase. The capture net should be larger than the fish, with a soft or padded rim. You can buy such nets with netting that is almost invisible in the water, making it much easier to drive the koi into them—just make sure you don't use an invisible net for the drive net!

Winter

UNDER THE ICE

What do you think the greatest danger to koi is in surviving a harsh winter? It isn't the cold. koi are able to tolerate freezing temperatures—sometimes even being very briefly frozen in ice—without damage. The primary cause of winterkill is, in fact, suffocation. Gas exchange is greatly reduced in an ice-locked pond, and as the ice grows thicker, so does the problem. The colder they get, the less oxygen your koi need, and they can survive with considerably less of it than they need during the summer, but they cannot survive if gas exchange is completely impeded.

In milder climates, where severe cold is not an issue, an excellent way of handling the gas exchange problem is to wait until the pond is completely frozen over. Chip a hole in the ice and pump out enough water to lower the level about an inch. You will now have an air space between the ice and the water, and the hole in the ice will permit air to flow freely, so there will be plenty of oxygen available for your fish. The air will also insulate the pond, making further freezing unlikely.

Where winters are colder, this obviously will not work, since the water will simply freeze solid under the original ice and air space. The best approach in this situation is to keep an area of the pond from freezing. Depending on weather conditions, this can be accomplished with paddles (either wind-driven or motorized) that keep the water in motion. It is surprising just how effectively this simple technique can prevent freeze-up. Electric heaters can also be used to keep an area ice-free. The idea is not to heat the pond, but simply to keep one area above freezing. As long as a section of the pond's surface is unfrozen, gas exchange can take place and your fish will not suffocate. Even under feet of ice, if there remains an interface between the liquid water and the atmosphere, there will be enough oxygen.

When keeping a hole open in the ice is not feasible, it is possible to bring air to the deeper waters mechanically. A compressor can deliver air to a bubbling device at the bottom of the pond. Because it is under pressure, the air will find its way out through the tiniest of fissures and holes, which would otherwise be inadequate to provide oxygenation of the water. You should start this aeration before the pond freezes solid to ensure that the air will be able to maintain exit holes.

DOWN IN THE DEPTHS

Despite the fact that fish have occasionally survived short periods being frozen in ice, they will not survive if their bodies freeze solid, so the only way to guarantee that your koi make it through the winter is to guarantee that there is always fluid water in the pond. Remember that the ideal koi pond has a very deep trough where the koi can go to escape both the heat of summer and the cold of winter.

WINTERIZING

Aside from making provisions for gas exchange under the ice, there are several preparatory chores that will help ensure the survival of your koi through the winter. First is a thorough cleaning of the pond. Nonhardy lilies should be removed to their winter locations. After the first light frosts and the koi's last feeding, the pump(s) should be turned off. Any lines or

other parts of the filtration system that need to be protected from freezing should be drained. Since your biofilter is going to need to be reestablished in the spring anyway, now is the time you should clean the biomedium. Remove accumulated gunk and grime and get everything ready so that in the spring all you have to do is reassemble it and turn on the power. The last thing needed is a massive water change so the fish do not have to deal with the stress from accumulated nitrates or other substances.

SPRINGTIME

Spring means that all the dormant organisms in your pond—from algae and protozoans to the koi themselves—become active again. What about the biofilter and its bacteria? Fortunately, a pond doesn't go from frozen to warm overnight. As spring approaches, life returns gradually to your koi pond. Once the danger of hard freezes is over, you should restart the filtration system. Even though you won't be feeding your fish yet, they will produce small amounts of waste as they become more active, and there will be accumulated nutrients from the winter. As long as the biofilter is getting a constant flow of oxygenated water, the bacteria colonies will regrow, and by the time the koi are feeding again and producing massive amounts of waste, the colonies will be at maximum populations. In essence, the biofilter will "cycle" right along with the pond, and by the time it is needed, it will be operational.

Index

Photo Credits

All-Japan NISHIKI-GOI Show, 27th Annual, 1, 4, 6, 7, 34, 35, 36, 37, 38, 39, 40, 41, 42, 43, 44, 45, 46, 47

Dr. H. R. Axelrod, 25

Nicholas Fletcher, 16

Michael Gilroy, 20, 21, 22, 26, 52, 55

Guido Lurquin, 30

MAG-NOY Israel, 56, 57, 61

Hugh Nicholas, 15, 51

MP. & C. Piednoir-Aqua Press, 9, 11, 17, 18, 31

Daniella Rizzo, 33

Stapeley Water Gardens, 13

Mary Sweeney, 24

Dr. Joseph L.Thimes, 32

John Tyson, 8, 10, 29, 53

Illustrations by David E. Boruchowitz